This Book Belongs to

_____

Copyright © 2019, Sherrita Berry-Pettus, M.Ed.

ISBN-13: 978-1-9453-4215-8

This book is sold subject to the condition that it shall not, by way of trade or otherwise, be lent, resold, hired out or otherwise circulated without the publisher's prior consent in any form of binding or cover other than that in which it is published and without a similar condition including this condition being imposed on the subsequent publisher. The moral right of the author has been asserted.

Illustrations Copyright © Sherrita Berry-Pettus, M.Ed.

Book Design and Illustrations by Uzuri Designs
http://uzuridesignsbooks.com

Hey Numbers!

By Sherrita Berry-Pettus, M.Ed.

Hey 1, is that you?

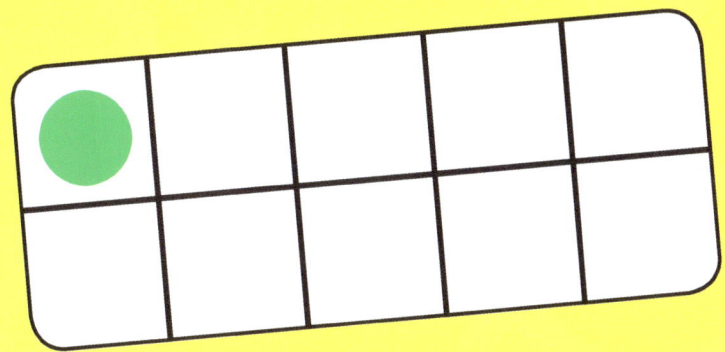

1

1, one, you're so fly!

6 7 8 9 10

2

Hey 2, is that you?

1 2 3 4 5

1, 2, two - a pair for you.

6 7 8 9 10

Hey 3, is that you?

3   III

3, three...1, 2, 3.

6 7 8 9 10

Hey 4, is that you?

1 2 3 4 5

5, five, count to 5.
1, 2, 3, 4, 5, I'm alive.

6 7 8 9 10

Hey 6, is that you?

1 2 3 4 5

6, six comes after 5.

6 7 8 9 10

Hey 7, is that you?

𝍷𝍷𝍷𝍷̸ 𝍷𝍷

1 2 3 4 5

7, seven comes before 8.

7

6 **7** 8 9 10

8, eight, looking like a snake.

6 7 **8** 9 10

Hey 9, is that you?

1 2 3 4 5

9, nine, you're so fine.

6 7 8 9 10

10, ten, and that's the end.
Let's count to 10, beginning with one.
1, 2, 3, 4, 5, 6, 7, 8, 9, 10!

6   7   8   9   10

Read the story again, whenever you want to **count to 10!**

www.ingramcontent.com/pod-product-compliance
Lightning Source LLC
Chambersburg PA
CBHW050749110526
44591CB00002B/30